NUMBER 573

THE ENGLISH EXPERIENCE

ITS RECORD IN EARLY PRINTED BOOKS
PUBLISHED IN FACSIMILE

(WILLIAM BRAITHWAITE)

GRANT TO W. BRAITHWAITE FOR PRINTING MUSICKE

n.p., (1636)

DA CAPO PRESS
THEATRVM ORBIS TERRARVM LTD.
AMSTERDAM 1973 NEW YORK

The publishers acknowledge their gratitude to
the Syndics of Cambridge University Library
for their permission to reproduce the
Library's copy, Shelmark: Dd*.3.27⁴(E)

S.T.C.No. 3540

Collation: A^4

Published in 1973 by

Theatrum Orbis Terrarum Ltd.,
O.Z. Voorburgwal 85, Amsterdam

&

Da Capo Press Inc.
- a subsidiary of Plenum Publishing Corporation -
277 West 17th Street, New York N.Y. 1011

Printed in the Netherlands

ISBN 90 221 0573 3

Library of Congress Catalog Card Number:
73-6105

HIS MAIESTIES GRACIOVS GRANT AND PRIVILEGE TO *WILLIAM Braithwait*, for the sole Printing, and publishing Musicke, *his way. &c.*

C*HARLES* by the grace of GOD, King of *England, Scotland, France*, and *Ireland*, defender of the Faith &c. To all, to whom these presents shall come, greeting. Whereas, we are informed by the humble petition of our well beloved subject, *William Braithwait*, Preacher, and Schoole-master. That after many yeares labours, studies, and endeavours, he hath accomplished an easie Method to facilitate Musicke, both for expedite, and exquisite composing; as also for speedy, and pleasant teaching, and learning it, by voice, or Instruments. For voice, that he singeth seven usuall numerall Monosyllables, one, two, three, foure, five, sax, siven, and expresseth them by seven

Arithmeticall figures. 1 2 3 4 5 6 7. diversly made, and that he singeth (as men reade) in one line, the name of the note appearing in it (tune and time, conspiring together in each figure) Which other Authors, and Writers have not heretofore done. And by the same petition he further sheweth, that he hath performed a *Tableture* for Instruments of Musicke, after the same manner, comprehending tune, and time in each Alphabeticall letter. And that also he hath devised to expresse the long, and short syllables in the *Greeke*, and *Latine* tongues, by the letters themselves, without the notes of length and shortnesse, which hath bin well approoved of, by sundry persons of learning, and judgment, and thereupon hath humbly besought us, in regard of his charges to be disbursed in publishing the said premisses, that we would be graciously pleased, to grant our Letters Patents of privilege unto him, for twenty one yeares, for the sole printing and putting to saile of any bookes, in the *Greeke* and *Latine* tongue, by his way so distinctly vowelled, and Musicke bookes, or lessons, by him so expressed as aforesaid. Know ye that we graciously favouring the industries of our subjects, in the investigation of such Art and Inventions, as may be of good use, and service to our people, and being willing to incourage them in all lawfull, and commendable studies, and endeavours, by appropriating unto them for some terme of yeares, the benefit and fruits of their inventions, of our especiall Grace, certaine knowledge, and meere motion,

tion, having given, and granted, and by these presents for us, our heires, and successors, doe give and grant unto the said *William Braithwait*, his Executours, Administratours & Assignes, full and free liberty, licence, power, privilege and authority, that He, his Executours, Administratours & Assignes, by him, and themselves, and by his, and their Deputies, shall and may, at all times hereafter, during the terme of twenty-one yeares, next ensuing the date of these presents, at his, and their owne proper cost, and charges, Imprint, and publish, or cause to be Imprinted, and published within this our Realme of *England*, and dominions thereof (according to the said Method, way and meanes, so by him devised) any booke or bookes, Poeme or Poems, Lesson or Lessons, for the more easie teaching, and attaining of Musicke, by voice, or Instruments, and for the furtherance of *Poetry*, *Oratory*, and gracefull pronuntiation of the *Greeke*, and *Latine* tongues, as aforesaid, and the same booke or bookes, Poeme or Poems, Lesson or Lessons, and other the premisses, being by him, or them, or his, or their deputy or deputies, so imprinted, or caused to be imprinted as aforesaid, to sell, utter, and put to saile within our said Realme of *England*, and dominions thereof, for his, and their best commodity and advantage. Wherefore our will and pleasure is, and by these presents of our more especiall grace, certaine knowledge, and meere motion, we doe for Vs, our Heires and Successours, straitly charge, and inhibite all and every person, and perſons whatſoever, other then the ſaid *William Braithwait*, his Executours, Adminiſtratours, Deputies and Aſſignes; that they, or any of them, doe not hereafter within this our Realme of *England*, or the dominions thereof, prInt, publiſh or put to ſale, or cauſe to be printed, publiſhed or put to ſale, any booke or bookes, Poem or Poems, Leſſon or Leſſons, after the wayes, Method, manner or formes ſo invented or deviſed by the ſaid *William Braithwait* as aforeſaid, nor ſhall import or bring, or cauſe to be imported or brought into this our ſaid Realme and Dominions thereof, from any the parts beyond the Seas, or from any our Realmes, or Dominions, any booke, or bookes, Poem, or Poems, Leſſon, or Leſſons, ſo imprinted as aforeſaid, during the ſaid terme of twenty-one yeares next after the date hereof, to be ſould in this our ſaid Realme and dominions thereof, without the licenſe, conſent and agreement of the ſaid *William Braithwait*, his Executors, Adminiſtrators & Aſſignes or ſome of them, firſt had and obtained in writing under his, or their hand, upon paine of the forfeiture, & loſſe of the ſame booke, or bookes, Poem or Poems, Leſſon, or Leſſons, ſo imprinted imported, and put to ſaile, contrary to the tenour and true meaning of this our preſent grant or privilege, and upon paine alſo of our high diſpleaſure, and ſuch other penalties and puniſhments as by the Lawes & Statutes of our Realme, or otherwiſe may be inflicted on the offenders therein: for their contempt, and neglect of our royall will, and

commandement

cõmandement, herein before declared. To Have and to hold the ſaid Licences, Powers, Privileges and Authorities, and other the premiſſes by theſe preſents granted, or mentioned to be granted, to the ſaid *W. Braithwait*, his Executours, Adminiſtrators and Aſſignes, for and during the ſaid terme of twenty-one yeares, from henceforth next enſuing, and fully to be complete and ended. And We doe by theſe preſents, for Vs, our Heires, and Succeſſours, give and grant unto the ſaid *William Braithwait*, his Executors, Adminiſtrators and Aſſignes, full power, and authority, that he, they, and every one of them, by him, and themſelves, or any of them, or by his or their Deputies, Servants, or Agents, ſhall and may, with the aſſiſtance of a Conſtable, or other lawfull Officer, at convenient time, and times, enter into any Houſe, Shop, Workehouſe or any other place or places whatſoever, where any cauſe of ſuſpition ſhall be had, and there to view and ſearch, aſwell for all ſuch booke, or bookes, Poem, or Poems, Leſſon or Leſſons, as ſhall be ſo printed, imported or brought into this our Realme, or Dominions thereof, to be ſold, as aforeſaid, as for all ſuch Muſicall-Arithmetical figures, Punchions, Matrices, Alphabeticall Letters for Tableture, and other ſuch tooles, and Inſtruments, as ſhall be uſed, had, made, or imployed, in or about the premiſſes, or any of them, contrary to the tenour, and intent of theſe preſents. And the ſame being found, to ſeize, and take, as forfeited to Vs, our heires, and ſucceſſors; the one moetie whereof to be to Vs our heires and ſucceſſors, the other moetie thereof tobe and remaine to the ſaid *William Braithwait*, his Executors and Aſſignes: And we doe by theſe preſents for Vs, our Heires, and Succeſſors, will, and command all, and ſingular Majors, Sheriffes, Iuſtices of Peace, Bailiffes, Conſtables, and Headborroughs, and other the Officers, Miniſters, and ſubjects of Vs, our Heires and Succeſſors, that they and every of them be from time to time during the terme aforeſaid, ayding, helping and aſſiſting to the ſaid *William Braithwait*, his Executours, Adminiſtratours, Deputies & Aſſignes in all things in and about the accompliſhment of our pleaſure herein declared, & in full, & due execution of theſe preſents, that they, nor any of them, doe at any time hinder, moleſt, or interrupt the ſaid *William Braithwait*, his Executors, Adminiſtrators, Deputies, or Aſſignes, or any of them, in the having, taking, or enjoying of the Licences, privileges, benefits and profits by Vs hereby granted and intended to him and them, as aforeſaid, as they tender our pleaſure, and will avoide the contrary at their perill: and theſe our Letters Patents or the inrowlment thereof ſhall be their ſufficient warrant in that behalfe. PROVIDED neverthelеſſe, and our will & pleaſure is, that if at any time during the ſaid terme of twenty-one yeares, it ſhall appeare unto Vs, our Heires or Succeſſours, or unto the Lords, and others of the privie Councell of Vs, our Heires or Succeſſours, that this our preſent grant is contrary to our lawes,

or

or in any ſort miſchievous, or inconvenient to the ſtate of this our Realme, then upon ſignification, and declaration to be made by Vs, our Heires, or Succeſſors, under our ſignet, or privie ſeale, or from the Lords, and others of the privie Counſell, or ſix or more of them (for the time being, of ſuch prejudice, & inconvenience as aforeſaid,) theſe our Letters Patents ſhall forthwith ceaſe, determin, and be utterly voide to all intents, conſtructions, and purpoſes. And wee doe hereby grant, for Vs, our Heires, and Succeſſors, that theſe our Letters Patents, or the inrowlment thereof, ſhall be good, firme, and effectuall in the Law, according to the true meaning of the ſame. NOTVVITHSTANDING the not mentioning or deſcribing, or the not certaine mentioning or deſcribing of the ſaid method, wayes, or meanes by the ſaid *William Braithwait* ſo deviſed as aforeſaid, and notwithſtanding any other defect, incertaintie, or imperfection in theſe preſents, or in any Law, Act, Statute, Ordinance, Proviſion, or Reſtriction whatſoever heretofore had, made, publiſhed, or provided to the contrary thereof in any wiſe notwithſtanding. Nevertheleſſe, our intent, and meaning is, that this our preſent grant of priviledge or any thing therein contained, ſhall not extend to prejudice any perſon, or perſons in his, or their lawfull proprietie, right, or intereſt for the ſole printing or publiſhing of any his or their Booke or Bookes, Poem or Poems, Leſſon or Leſſons, Copie or Copies, ſo as he, or they doe not print, or publiſh the ſame according to the method, way, or invention of the

ſayd

ſayd *William Braithwait*, any thing in theſe preſents contained to the contrary in any wiſe notwithſtanding. ALTHOVGH expreſſe mentiō of the certaintie of the premiſſes, or any of them, or of any other gift, or grant by Vs, or any of our Progenitours or Predeceſſours to the ſaid *William Braithwait* hertofore made in theſe preſents, is not made or any Statute, Act, or Ordinance, Proviſion, Proclamation, or Reſtraint heretofore had, made, ordained or provided, or any other thing, cauſe, or matter whatſoever to the contrary thereof in any wiſe notwithſtanding. In witneſſe whereof wee have cauſed theſe our Letters to be made Patents. Witneſſe our ſelfe at *Canbury* the eighteenth day of *Auguſt* in the Eleventh yeere of our Reigne.